Manipulating Science for the
Promotion of Viagra

Evidence from the Medical Literature

Manipulating Science for the Promotion of Viagra

Evidence from the Medical Literature

Dominika A. Boczula

LULU PRESS, MORRISVILLE, NORTH CAROLINA 27560

13 12 11 10 09 08 07 06 05   2 3 4 5

ISBN 978-1-4357-1046-7

Library of Congress Cataloging-in-Publication Data

Boczula, Dominika A.

Manipulating science for the promotion of Viagra: Evidence from the
medical literature/
Dominika A. Boczula

p.; cm.
Includes bibliographical references and index.
ISBN 978-1-4357-1046-7   (paperback: alk. paper)
    1.   Pharmaceuticals 2. Medical Sociology

[DNLM: 1. Drug Marketing.   2. Drug Promotion. 3. Sociology.
RC889 .B64]   I. Title.

R727.3. B66 2008

610.69'6-dc21

The paper used in this publication meets the minimum requirements of
the American National Standard for Information Sciences –
Permanence of Paper for Printed Library Materials, ANSI Z39.48-1992.

Studying up does not mean being submitted to the agenda of those we study: what some disgruntled scientists conclude from our research remains their business, not ours. As far as I can tell...they might have concluded that the white purity of science should never be sullied by the dark and greasy fingers of mere sociologists.

Bruno Latour,
Reassembling the Social

Acknowledgement

I would like to thank my family, friends and close ones in Canada and abroad for their support during the writing and publication of this book, as in all other times of happiness and tribulation. This could not have been done without their support or that of James Jegen. Comments and suggestions provided by Dr. Ariel Ducey on earlier drafts of the manuscript are also kindly appreciated.

Contents

Introduction

In his book entitled *Science in Action*, the sociologist Bruno Latour attempts to "penetrate science from the outside, following controversies and accompanying scientists up to the end, being slowly led out of science in the making" (1987). I too would like to employ this method in order to explore and expose 'science in the making', albeit confined to a more specific genre. Latour notes that "apart from those who make science, who study it, who defend it or who submit to it, there exist, fortunately, a few people, either trained as scientists or not, who open the black boxes [facts] so that outsiders may have a glimpse at them." I would like to open one of these 'black boxes' by focusing on the pharmaceutical industry's, particularly drug manufacturer Pfizer Pharmaceuticals', mobilization of science and scientists towards the construction of the 'fact' of erectile dysfunction. From my research, it becomes evident that so-called facts believed by many can become seen as such based on the decisions and tactics made by a powerful few.

In formulating my research questions, I decided to explore the advancement of medicine into the realm of sex, and more specifically into the realm of erectile function. I also wanted to illustrate how Pfizer successfully created a market for Viagra using the medical literature, as well as to find who is involved in the production and marketing of Viagra, and how this is done. Arnold Relman, a critic of the pharmaceutical industry's relations with the general medical profession, is well-known for his theory on the medical industrial complex. Although he examines arrangements in today's "market-oriented, profit-driven health care industry" such as continuing medical education for physicians provided by corporations, I would like to extend his analysis by looking at a specific relationship between the health care industry and a specific pharmaceutical (1991). Through my research, I hope to examine, and contribute to this topic with regards to whether there is evidence of a medical industrial complex occurring in the case of Pfizer's Viagra. Lastly, I wanted to make all of the above inquiries from a sociological perspective, a perspective which is evidently lacking in the literature on the subject.

Multiple factors influence the adoption of a new medical treatment into medical practice. These include the previous difficulty in treatment of the condition, the facility and risks associated with the new treatment, information provided by sales representatives and the media to both health care providers and the public, as well as perceived patient demand by health care providers (Harrold *et al.*, 2000). I argue that with these points in mind, individuals at Pfizer set out to reinforce, downplay, or construct facts and ideas in order to promote the sales of Viagra. I am not attempting to overlook or deny the social and economic factors which enabled the rise of the medical problem known as erectile dysfunction. Rather, I argue that many of these factors were skillfully employed by one group of entrepreneurs: the marketing force at Pfizer.

My examination of the evidence provided by the medical literature reveals a powerful corporation's transformation of a bodily condition into a legitimate disease, and a drug into the legitimate treatment of this 'disease' through many skillful, yet manipulative moves. The well-contrived financial support of certain researchers, positioning of certain research findings, relegation of side effects, as well as the intentional empowerment of patients while overshadowing the medical profession, were and are strategies which worked to make Viagra the fastest selling prescription drug in the history of the United States, and in the process helped to promote the medicalization of a problem with larger social ties. My focus centers on the people who constructed the product and the demand. They include more than just individuals at Pfizer Pharmaceuticals, but also researchers, physicians, consultants, and patients, all of whom contributed in large part to make Viagra into the product and phenomenon, which it has evolved into.

Methodology

While designing a project to examine the sociological aspects of Viagra, I decided to focus on the mobilization and manipulation of science that had to occur in order for Pfizer's 'dream' to develop into the phenomenon which it has become. I set out hoping to enlighten about the multitude and complexity of interdependent factors involved in the creation of the brand. I employed largely qualitative methods in order to perform a close, in-depth study of social groups invested in the Viagra phenomenon, as well as to closely track the evolution of medical, historical and social dynamics over time.

Medline, an electronic search engine which indexes articles from 1966 onwards, was used as the main source of data on articles and authors relating to sildenafil publishing. The database was first searched using the search term sildenafil.ti, which identified articles in which sildenafil was featured in the title. The search was also limited to 'human' and 'journal article', resulting in a set of 745 articles. The data set was downloaded, cleaned and imported into a Microsoft Excel spreadsheet. This comprised the database. This database was then supplemented with select international articles. I performed an in-depth content analysis of this select sample of international journal articles, as well as of major medical journals' comments, letters and editorials.

The Science Citation Index was also employed. This tool is a multidisciplinary index to the journal literature of the sciences. It fully indexes 5,900 major journals across 150 scientific disciplines, while also including all cited references captured from indexed articles. A search of all articles published from 1975 to 2006 in the Science Citation Index with sildenafil in the title, abstract, or keywords, not limited by language or article type, resulted in 2500 articles. Using these articles, cross-tabulations were created to indicate the 100 most common authors within the 500 most frequently cited articles on sildenafil, the 100 most cited articles, the 100 most common journal titles, and the top 34 countries publishing articles on the topic.

Viagra's Introduction: the Context

Those who set and organize allopathic medical school curricula, medical governing boards, and medical discourses are undoubtedly conferred power, autonomy and control over scientific and medical knowledge, practices, and their practitioners and consumers. The ruling body's decisions are not only guided by evidence and objective truths. They are also influenced by values, contexts, lobby groups, funding, and pressures from other individuals and organizations. In addition, medical power not only resides in institutions or elite individuals, but is also deployed by every individual by way of socialization to accept certain values and norms of behavior (Lupton, 2003). We understand and refer to the body in ways which reflect the dominant discourse of the time.

It has been suggested that the public's tolerance for mild symptoms and benign problems has decreased, spurring a progressive medicalization of physical distress in which simply uncomfortable states and isolated symptoms are reclassified as diseases. Many conditions, such as alcoholism, obesity and hyperactivity have recently been categorized as diseases, a label which conveys assumptions about culpability.

Medicalization occurs when previously non-medical problems are defined and treated as such, usually in terms of illnesses or disorders. While medicalization can be bidirectional, there is strong evidence for increases in medicalization over the past three decades (Conrad and Leiter, 2004). Today, as a result of changes in marketing and deregulation, high profitable "blockbuster" lifestyle medications, which improve quality of life, but not necessarily health, have proliferated and are now included under the broad term of medicine. Ironically, by promoting new standards of health and well-being and by playing on insecurities, pharmaceutical companies can expand their markets infinitely. In doing so, they increasingly medicalize discontent.

For example, the inability to obtain an erect penis by some men has been known to exist for decades, and presumably much longer. Such a physical condition has undoubtedly interfered with procreation and intimate relations, and as such, individuals have been seeking aids to improve sexual performance and enhance fertility for centuries. Soderling and Beavo note that although not life-threatening, the psychological and social consequences of this condition are serious as well (2000).

Erectile function in men depends upon a complex interaction of psychogenic, neurologic, hormonal and vascular factors, and the management of erectile dysfunction would ideally reflect this complexity of control. Therapeutic options include psychological and non-pharmacological approaches such as counseling for interpersonal difficulties or addressing lifestyle factors that contribute to erectile dysfunction such as cigarette smoking or alcohol abuse. However, despite the frequent involvement of emotional and interpersonal factors in sexual dysfunction, medical treatments are often viewed as more efficient and effective, and as a result, preferred over other treatments (Levine, 1992). The treatment for erectile dysfunction followed this trend when the general public, notably men, began turning to the medical field for a way to combat this "side effect of socially rooted problems" through allopathic means (Carpiano, 2001). By 1994, Tiefer noted that (preferred) forms of ED treatment had indeed moved away from psychogenic causes in favor of organic ones such as penile, surgery, implants and injections, although their results were mixed.

Two terms have been largely employed to label this condition in men: impotence and erectile dysfunction. Both terms denote similar, yet distinct concepts. The term impotence has traditionally been used to signify the inability of a male to attain and maintain erection of the penis sufficient to permit satisfactory sexual intercourse. This value-laden label, which means 'without power' in the Latin language, symbolizes a fault with the man himself for the condition and captures the tendency to blame. A man termed 'impotent' is devalued as one no longer able to fulfill his role as a 'true' man in society. It does not however, hint of reasons for the inability to attain an erection, which may be truly outside of the control of the man. Conversely, the term erectile dysfunction is used to signify an inability of the male to achieve an erect penis as part of the overall multifaceted process of male sexual function. This process comprises a variety of physical aspects with important psychological and behavioral overtones (National Institute of Health Consensus Development Panel on Impotence, 1993). However, being labeled as having this condition, as opposed to impotence, is less likely to impinge on one's sense of value as a man or ability to function as a 'true' man in society.

The development of this new term of erectile dysfunction and the construction of the condition which it denotes helped to transform unacceptable erectile performance into a subject for medical analysis and treatment, and as I argue, in the process further blurred the boundary between discontentment and disease. In addition, by analyzing the events which are described below through the lens of Latour's theory of fact-construction, it becomes evident that erectile dysfunction became an actual condition, a fact, rather than simply an obscure term. In order to transform the term from one of insignificance into one which was recognizable and accepted, many events had to take place. These events included Pfizer's construction and 'branding' of erectile dysfunction, the creation and incessant marketing of a link which made the condition synonymous with Viagra, and the transformation of previously embarrassing events occurring in men's bodies into a recognized and accepted universal 'fact'.

The term erectile dysfunction was first used in the literature by Levine in 1976 in order to describe impotence originating from mixed organic and psychological causes. It has been argued however, that the Pfizer-sponsored Massachusetts Male Aging Study (MMAS), the results of which were published in 1994, actually put the term 'on the map' (Loe, 2004; pg. 48). In this important study for the field of urology, impotence was assigned a new name and redefined more broadly. The MMAS was a subjective, self-administered questionnaire characterizing erectile potency not as an either/or but rather as a continuum. On the questionnaire, subjects were asked to rate their potency on a scale of one to four: (1) not impotent, (2) minimally impotent, (3) moderately impotent, or (4) completely impotent. These responses were then assigned gradations of erectile dysfunction, ranging from "no ED" to "mild ED (usually able)",

to "moderate ED (sometimes able)" to "severe ED (never able)".
Those men who reported "unsatisfactory sexual performance" or
who were "usually able to penetrate partner" were included in
the "mild ED" category. Given these flexible definitions, of 1
290 men surveyed aged forty to seventy, 52% fell in the "mild,
moderate, or complete erectile dysfunction" categories.

This statistic, that half of men over forty experience ED,
was published in the *Journal of Urology* and is now cited
regularly by Pfizer in their promotional material. Based on these
figures, Dr. Irwin Goldstein, a prominent urologist working as a
consultant for Pfizer, estimated that the number of American
males "impaired" by sexual dysfunction was thirty million,
thereby creating a huge potential market for oral sildenafil (Loe,
1994). This finding also found its way onto Pfizer's promotional
materials which often state that "ED, to some degree, affects
approximately 30 million males in the United States" (Pfizer
Pharmaceutical Website, 2006). The incidence of the condition
is known to increase with age, and estimates by other sources
confirm that the number of American men with erectile
dysfunction ranges from 10 to 20 million. Inclusion of
individuals with partial erectile dysfunction increases the
estimate to about 30 million (NIH Consensus Development
Panel on Impotence, 1993; Feldman *et al.*, 1994; Kinsey *et al.*,
1948). A majority of these individuals are older than 65 years of
age, with a prevalence of about 5% observed at age 40,
increasing to 15-25% at age 65 and older (NIH Consensus
Development Panel on Impotence, 1993).

By supporting the development and promotion of the
findings from the MMAS, Pfizer largely constructed erectile
dysfunction in terms of 'who has it'. In order to answer what
erectile dysfunction actually is, the corporation decided to make
the inability to get an erection solely that - a mechanical problem
stripped of the important psychological and behavioral overtones
which were once a crucial part of its definition. This was done
largely through funding the creation of the International Index of
Erectile Function (IIEF). The index is a subjective measure
composed of 15 questions, focusing solely on penetration as a
measure of sexual satisfaction. It is used extensively by Pfizer in
publishing, including by Goldstein *et al.*, who used two of its

questions as the main measures of sildenafil efficacy in the first peer-reviewed article on sildenafil, which is also the most cited (Goldstein *et al.*, 1998). Although the IIEF is often stated as a means of treatment analysis by researchers, upon closer examination of research methods, it becomes apparent that only selected questions from the IIEF are often used to evaluate subjects' results. In theory, it is possible to publish only the results from questions which support one's hypothesis, and not those from all fifteen.

Perhaps presenting physical symptoms of erectile dysfunction does not mean that one is "suffering in silence" (Pfizer pamphlet, 2000). In their study evaluating the epidemiology of male sexuality in Germany, Braun and colleagues found that when treatment need was defined as the co-occurrence of erectile dysfunction along with dissatisfaction with one's sex life, 6.9% of men were determined to be in need of erectile dysfunction treatment, which is significantly less than any estimated prevalence of the condition itself (2000). In a related study, Klotz *et al.* found that the need for erectile dysfunction treatment is modulated by numerous outside factors such as situational and partner-related reasons (2004). These findings may also hint as to why the abandonment rate of Viagra is relatively high.

From Figure 1, it is evident that the term impotent remained more popular in the scientific literature until the late 1980s, after which the term erectile dysfunction gained popularity. From the figure, two trends are noted: first, the entire field of erectile function became more prevalent in the literature beginning in the late 1980s, and also, that with time, the term erectile dysfunction became far more common than the previously-used terms. The usage of the term erectile dysfunction in the literature grew exponentially in the 1990s, possibly due to the NIH Consensus Development Panel's preference for the term, the MMAS study's usage of the term, or Pfizer's fondness for, and frequent employment of the term in sponsored research.

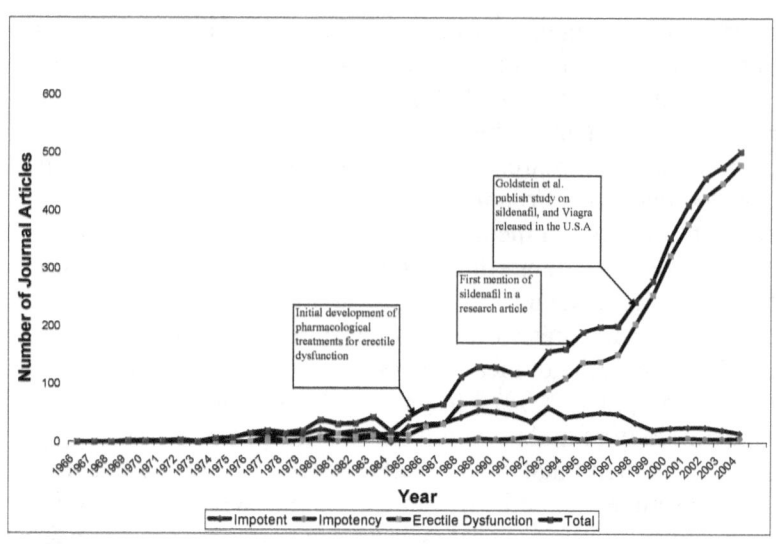

Figure 1 – Results of a Medline search of journal articles containing each respective term in one or more of the following: the title, abstract, 'name of substance' word and 'subject heading' word, by year.

Outsmarting the Mainstream Medical Profession

Initially tested as a heart medication due to its vessel-dilating properties, sildenafil citrate was found to have unanticipated repercussions in male trial subjects: it effectuated erections. The compound's manufacturer, Pfizer Pharmaceuticals, named the compound Viagra and evaluated its efficacy in numerous randomized, placebo-controlled trials involving more than 3000 men with varying degrees of impotence associated with diabetes, spinal cord injury, history of prostate surgery as well as those with no identifiable organic causes of ED (United States Food and Drug Administration Website, 1998). Unlike previously approved treatments for erectile dysfunction, Viagra was not found to cause penile erections directly, but to instead affect the man's response to sexual stimulation. By enhancing the smooth muscle relaxant effects of nitric oxide, a chemical normally released in response to sexual stimulation, the drug enhanced smooth muscle relaxation which allowed for increased blood flow into certain areas of the penis, leading to an erection.

The FDA announced the approval of Viagra on March 27, 1998, while Health Canada's approval came on March 8, 1999. In the United States, Viagra quickly became that country's best-selling prescription drug ever, with 7 million prescriptions written by mid-December of the same year (*Canadian Medical Association Journal*, 1999). The initial approval of Viagra in the United States, as well as its perceived ability to 'restore' malfunctioning masculinities, resulted in unprecedented demand for the drug internationally (Potts *et al.*, 2004). In 1998, a road connecting Canada with upper New York State was dubbed the "Viagra Highway" due to the number of Canadians crossing the border to buy the drug (*Canadian Medical Association Journal*, 1999).

Within a few months of its release, Viagra and the phrase 'male erectile dysfunction' became household terms, penetrating news and multimillion dollar advertising campaigns, allowing the previously hidden, taboo-like health and social problem to surface along with its accompanying questions. Men were made to feel that it was acceptable and common to have this 'dysfunction', especially since they would no longer have to "suffer in silence", seemingly alone:

Erection problems are very common, and are now understood to be a real medical condition that can be treated. Many men are suffering in silence when all they need to do is talk to their physician. (Pfizer pamphlet, 2000)

The marketing campaign and efficacy of Viagra propelled its sales to 50,000 prescriptions filled per month in Canada in 2004. In addition, within five years of Viagra's launch, the drugs Cialis and Levitra, both working similarly to Viagra, also entered the men's market. In 1997, the American Urological Association's newsletter Urology Times forecast that urologists would see a new field of medicine "explode" in the near future (Loe, 2004). This undoubtedly came to fruition, with these advancements creating a giant new product category for the pharmaceutical industry – nicknamed the 'sexuopharmeceuticals' (Tsao, 2004; Potts *et al.* 2004). During this time period however, the previously changing balance of power between physicians, consumers and pharmaceutical corporations became not only evident, but also exploited by both consumers and the pharmaceutical industry.

The approval of Viagra was accompanied by extensively advertised claims that large numbers of men were suffering from erectile difficulties and seeking treatment. In media campaigns and advertisements, the slogan "talk to your doctor" hinted of Pfizer's continual encouragement of men to visit their physicians in order to inquire about ED and its accompanying treatment. After all, a man in a medical office inquiring about Viagra is more likely to receive a prescription than one who does not address the issue at all. This approach centred on delivering potential customers to the 'gate-keepers' of Viagra sales, the physicians, who ironically also stood to profit from such urgings (Conrad and Potter, 2000).

In their 2000 study, Harrold *et al.* examined the use of Viagra in the first six months of its availability in a managed care setting in order to determine prescribing trends, characteristics of Viagra users, prescriber characteristics, and usage patterns in a cohort of Viagra users. They found that media attention clearly resulted in a greater awareness of erectile dysfunction as a treatable condition, and increased the demand for the drug. In fact, almost 60% of American patients who were prescribed Viagra had never sought medical attention for the problem, and as such had no documentation of prior treatment for erectile dysfunction. The study also found that 85% of first-time prescriptions being filled for the drug occurred within the first twelve weeks of its availability (Harrold *et al.*, 2000).

Physicians have always been the direct link between pharmaceutical producers and patients, which is why the pharmaceutical industry spends billions of dollars on symposia and galas for physicians, offering incentives for prescriptions, and advertising and promoting their products to physicians. The role of physicians as "providers" however, is changing in the current medical marketplace, particularly due to direct-to-consumer advertising undermining their authority with regards to which drugs to prescribe (Conrad and Leiter, 2004). The Federal (USA) Drug Administration Modernization Act of 1997 made drug advertising both more feasible and more attractive to pharmaceutical manufacturers, a development which Pfizer used to their advantage when marketing Viagra in 1998 and onwards. The changes to the act, which now allowed television and radio advertisements to name both the disorder and the drug's benefits without a lengthy summary of potential side effects and contraindications, are seen as the main reason for annual spending on direct-to-consumer advertising for prescription drugs tripling between 1996 and 2000 (Conrad and Leiter, 2004).

Pharmaceutical companies claim that direct-to-consumer advertising has an educational function that creates better-informed consumers, encouraging them to consult their physicians about underdiagnosed symptoms and treatment options, and enabling patients to make better choices with regards to their health care (Bonaccorso and Sturchio, 2002; Lyles, 2002). The American College of Physicians has stated its position that consumer advertising "does not constitute appropriate patient education" (Maguire, 1999). Regardless, 3.5% of Canadian patients, and 8.2% in the United States, report using advertising as an information source (Mintzes *et al.*, 2003). Although these figures appear negligible, it must be considered that although some patients may not consciously employ such sources to provide information, they nonetheless internalize the messages they are exposed to daily by various media.

A study by Mintzes *et al.* compared prescribing decisions in a US setting with legal direct-to-consumer advertising and a Canadian setting where such advertising of prescription drugs is illegal, but some cross-border exposure occurs (2003). The results suggest that more advertising leads to more requests for advertised medicines, and more prescriptions. If direct-to-consumer advertising opens a conversation between patients and physicians, that conversation is highly likely to end with a prescription, often despite physician ambivalence about treatment choice. Pfizer's advertising of "ask your doctor" takes advantage of such dynamics between patients and physicians. Given that a potential patient inquires about Viagra to their physician, a prescription, and concurrent sale of the drug is more likely than not. According to Mintzes and colleagues, physicians fulfilled, on average, 75% of requests for direct-to-consumer advertised drugs (2003). Conrad and Leiter hypothesized that pharmaceutical manufacturers are circumventing physicians' control over knowledge regarding available drugs (2004). The case of Viagra certainly supports this notion.

Physicians are increasingly frustrated that the developments associated with increased direct-to-patient advertising are putting their patients in the "diagnostic driver's seat" (Maguire, 1999). Some note that increasingly, patients are presenting them with lists of drugs which they would like to try, many of which are neither "time-tested" nor "cost effective" (Maguire, 1999). Other physicians state that patients believe that certain (advertised) drugs are going to be a panacea for their problems, and as a result pressure them for prescriptions, regardless if the physician feels otherwise. It must be noted however, that these sentiments rely on the belief that physicians are selfless workers on the patient's behalf who only do what is necessary for them, including prescribing only time-tested and cost-effective medications. If only this was always true, then pharmaceutical companies, through direct-to-patient advertising, would be exploring a new method of increasing sales: through the obtrusion of physician-patient relations. This method however, is not new, but rather, has been employed by drug manufacturers for some time.

Although, pharmaceutical companies are investing large amounts of resources into consumer advertising, they are also spending increasing amounts on convincing physicians to prescribe their products: while half a billion dollars was spent on television commercials during the first nine months of 1998, five times as much, or $2.7 billion was spent on sales and promotional efforts to office-based physicians (Maguire, 1999). With regards to Viagra alone, these efforts concentrate on keeping doctors updated on success stories, educated about sexual dysfunction, and stocked with information sheets, free Viagra samples, educational videos, and other Pfizer paraphernalia (Loe, 2004). After all, at nearly $10 per pill, physicians still serve as the official gatekeepers to lucrative profits from sales of Viagra for pharmaceutical companies.

With regards to the traditional method of disseminating drug-related information about new treatments to medical professionals, individuals at Pfizer decided to exercise a more powerful approach. This occurred by financially supporting key researchers and publications contributing to the universe of knowledge relating to erectile dysfunction. As such, the dispersion of scientific information regarding erectile dysfunction and its treatment by Viagra was to an extent skewed towards the benefit of Pfizer. In addition, a large volume of published literature regarding Viagra, especially before its approval and immediately after, created a false sense of importance and interest on the drug, which was eventually, termed 'The Viagra Phenomenon'.

It is no secret that pharmaceutical manufacturers, as others in industries involved in the development of products, often fund researchers, their projects, symposia and certain professional gatherings. In return, certain physicians "moonlight as consultants" for these companies (Stipp *et al.*, 1998). Experts and marketers are increasingly dependent upon one another, as experts need funding to research, and marketers need expert status to achieve legitimacy and to publish data. This arrangement constitutes a large portion of the modern medical landscape. Katherine Greider, author of the book entitled *The Big Fix: How the Pharmaceutical Industry Rips off American Consumers*, writes that this fusion of science and capitalism has left America (as can be assumed for Canada as well) "oddly impoverished in the way of unbiased, approachable information about the usefulness and cost of one drug versus another" (2003; pg. 2). Support for this statement is found in a disturbing study in *The New England Journal of Medicine, which* revealed that 96% of medical experts who published studies or other articles supporting the use of certain controversial blood-pressure drugs had financial ties with companies that make them, while only 37% of those who wrote articles critical of the drugs had such ties.

The practice of stating one's affiliations, potential conflicts of interest, and sources of research funding when submitting a publication serves as one method of acknowledging such links. However, although a journal such as *Current Medical Research and Opinion* requires that authors affirm that the publication of their article is free of any conflicting commercial interest, and that no commercial funding is associated with the given paper, the affiliation or interests of the authors are undisclosed. Often, Pfizer's support is disclosed and acknowledged by authors. Aside from such instances where funding and professional relationships are overtly disclosed, however, the corporation also supports certain influential projects cryptically.

The Importance of Viagra: Concurrence vs. Dissension

In the age of medical progress, scientific knowledge and medical answers are generally unquestioned as the best, most efficient, most legitimate solutions. However, the history of science, medicine, and technology is also a history of attempting to solve social problems and control populations.

-Meika Loe, author of *The Rise of Viagra: How the Little Blue Pill Changed Sex in America*

In the late 1990s, Viagra emerged as a pop culture phenomenon due to the massive public response and plethora of international attention given to the drug by all modes of the media. It seemed that every publication, radio and television station opted to add the drug to its assembly of newsworthy items, albeit generally focusing on a specific aspect relating to its efficacy, development, safety, or social impact. Viewed optimistically, the barrage of media attention also served to legitimize sexual dysfunction, bringing the topic "out of the closet" as never before (Rosen, 1998).

Hence, Viagra, considered such a popular 'news-worthy' event worldwide, as well as a completely new development in the field of erectile dysfunction treatments due to its reliability, discreet administration and minimal side effects (Goldstein *et al.*, 1998), would undoubtedly capture the attention of even the most prestigious medical journals. However, the attention bestowed on the drug by these publications was lukewarm at best. I propose that a controversy arose in the international research community largely 'under the radar' at the time of Viagra's launch, as well as in the coming years.

Clinical medicine and nursing titles comprise the majority of a typical hospital library's journal list and are of critical importance to large academic medical and research centers. Selected English language core clinical journals are found in the *Abridged Index Medicus,* which was published by the National Library of Medicine (NLM) from 1970 - 1997. Although the NLM no longer produces the *Abridged Index Medicus* as a print publication, the value of the titles as a core list of "selected titles of biomedical journal literature of immediate interest to the practicing physician" is still recognized (United States National Library of Medicine, 2005). These titles continue to be searchable on NLM's Medline database as a search subset limit called 'core clinical journals'. Two of the most prestigious core clinical journals are titled *The Lancet* (published in England), and the *Journal of the American Medical Association* (JAMA).

If the complete archives of *The Lancet* are examined, a surprising find transpires. Of the 25 articles mentioning sildenafil in the title, abstract or keywords, only 7 are journal articles. The others are news events, commentaries, editorials and letters. Of these 7 journal articles, only 3 address sildenafil as a treatment for erectile dysfunction, including one focusing solely on its retinal side-effects. The remaining 4 journal articles report new, non-ED related, clinical uses for sildenafil. These include its use as a treatment for lung fibrosis and pulmonary hypertension as well as its effect on local inflammation reactions experienced by Crohn's disease patients. From the examination of this journal, it becomes evident that sildenafil is seen as an interesting compound, but not necessarily as a compound of interest solely for the treatment of erectile dysfunction. An examination of the archives of JAMA nets similar findings. Of the 21 articles mentioning sildenafil in the title, abstract or keywords, only 7 are journal articles. All 7 of these journal articles, however, address sildenafil as a treatment for erectile dysfunction. If the term 'Viagra' is searched in these archives, although more items such as news reports and commentaries are found, none of the search results are journal articles.

The fact that these core clinical journals did not publish many articles on sildenafil may suggest that the treatment of erectile dysfunction with Viagra was not, and is not seen as a very legitimate medical concern in the mainstream medical community. Debates in the letters, editorials, and commentaries about whether ED is a serious problem and one that the medical profession should be paying attention to, highlight this stance.

For example, within two months of Viagra's approval in the United States, *The Lancet* published a news feature entitled 'Is the honeymoon over for Viagra?', in which the author paints a frightening image of Viagra as having already been confirmed by the FDA as being linked to the deaths of six men in the United States, leaving three men in Egypt in need of intensive care after its use, and causing the Israeli health ministry to prohibit its physicians from prescribing the drug.

The same journal published an editorial in September of 1998 in which the relatively easy availability of the drug is criticized. The editorial states that "the speed of transfer of this agent from an FDA license to a drug of misuse has been remarkable but that is no reason to give up on attempts to control distribution". It continues to state that the manufacturer will want to cooperate with drug regulators in bringing illicit sources and distribution networks under control, however difficult this might be, and concerns about the impact on health services indicate that restriction, for the time being, to hospital prescription only would be sensible for Europe. Two months later, *The Lancet* published another article cautioning physicians of potential issues with the drug. The article states that "sildenafil is no panacea" and that not all patients with erectile dysfunction will benefit from it (Chan-Tack, 1998). It continues that there is substantial risk that many patients will receive little or no evaluation before treatment, the potential for ill-informed and inappropriate prescribing is high, that sildenafil is not an aphrodisiac, does not increase sexual desire or libido and has a high potential for abuse by thrill-seekers. The article also notes that the known side-effects may not be transient, "as current data suggest" (Chan-Tack, 1998). It continues that, "sildenafil may also have other, as-yet-unknown, adverse effects that will become evident only over time" (Chan-Tack, 1998).

The *Journal of the American Medical Association*, although not as unenthusiastic of sildenafil therapy as *The Lancet*, in its commentaries, editorials and news features also tends to advise physicians to prescribe sildenafil cautiously. In various issues, physicians are advised to exercise extreme caution when prescribing sildenafil to men with diabetes, or those with coronary artery disease. Physicians are also made aware of the possibility of hypotensive reactions in patients taking antihypertensive drugs and sildenafil and are instructed to alert their patients about this potential adverse effect.

So, if the core medical journals were not publishing articles on sildenafil, then who was? In our database, *The International Journal of Impotence Research* is the most common journal title, having published 53, or 7.1% of the 745 articles. These findings are also supported by those from the *Science Citation Index*. The *International Journal of Impotence Research* was established eighteen years ago as an offshoot of the Nature publishing group, and lists no official sponsors on its website. However, the editor-in-chief, as well as the remainder of the editorial board have overwhelming financial relationships with Pfizer. Of the nine members, seven serve as speakers, consultants, scholars, researchers or recipients of research funding from Pfizer, all in multiple categories. A qualitative exploration of the articles published by the journal reveals a decidedly positive view of the drug, with minimal mention of side effects or unfavorable research results. This may stem from the fact that investigators and consultants for Pfizer must generally sign nondisclosure agreements that prevent them from divulging data that might conflict with the company's reports (Loe, 2004).

An examination of the top 100 most common authors publishing on sildenafil indicates a similar level of sponsorship. Of the top ten authors, eight openly receive research support from Pfizer, including all of those in the top six. The most common author within these 500 articles is Dr. R. Kloner, a urologist who serves on the *International Journal of Impotence Research* editorial board, as a speaker and consultant for Pfizer, and who also co-authored the book entitled *Viagra: How the Miracle Drug Happened and What it Can Do for You! (1998)*.

The second most common author is Dr. I Goldstein. As a Boston University urologist who also serves as consultant and spokesperson for Pfizer, Dr. Goldstein was "solely entrusted with the 'branding' of ED and teaching doctors and the public at large about ED" from the 1990s onwards, largely constructing and revealing the necessity for an impotence drug such as oral sildenafil (Loe, 2004). As mentioned above, this was done largely by developing and publishing the results from the MMAS, 'estimating' the extent of the 'disease' of erectile dysfunction in North America, and serving as chief author of the first peer-reviewed paper outlining sildenafil's use as a treatment for erectile dysfunction. While appearing in media venues everywhere from *Playboy* to ABC's *Good Morning America* to tout Viagra as "a dream practitioners in this field didn't think possible" and "the start of an exciting revolution", Dr. Goldstein did not readily divulge affiliations with Pfizer, thus appearing to the public as an unbiased expert (Loe, 2004).

Interesting findings also arose from the search of the *Science Citation Index* with regards to the top 100 most cited articles with sildenafil in the title, abstract, or keywords. Of the top 25 articles, more than half have at least one author who publicly receives funding from, or serves as a consultant or speaker for Pfizer. It also becomes apparent from an exploration of this group of articles that aside from funding individual researchers, Pfizer also takes the more direct route to disseminate much-needed studies on Viagra: the formation of their own research teams.

Although the first article to appear about sildenafil in 1996 already associated it as a treatment for erectile dysfunction (completed at Pfizer Central Research in the U.K.), it serves as the third most cited article about the drug (Boolell *et al.*, 1996). The second most cited study focuses on drug biochemical properties, more specifically in reviewing recent work on the newly discovered cyclic nucleotide phosphodiesterase (PDE) families, as well as other specific PDEs in the regulation of T-cell activation, insulin secretion, growth, fertility and penile erection (Soderling and Beavo, 2000). With such a broad range of conditions and treatments related to PDEs, the article by Soderling and Beavo is undoubtedly cited in many publications which are not related to erectile dysfunction or sildenafil usage. Hence, it appears as the second most cited article associated with sildenafil almost by chance.

The most cited article was completed by Goldstein *et al.*, as part of the Sildenafil Study Group's study on the evaluation of the efficacy and safety of sildenafil (1998). The article details three studies performed on American men: a 24-

week dose-response study of 532 men treated with oral sildenafil or placebo, a 12-week flexible dose-escalation study of 329 men treated with oral sildenafil or placebo, as well as a 32-week dose-escalation, open-label extension study of 225 of the 329 men. In all three of the studies, men were excluded if they had penile anatomical defects, a primary diagnosis of another sexual disorder (ex. premature ejaculation), spinal cord injury, any major psychiatric disorder not well controlled with treatment, poorly controlled diabetes mellitus, active peptic ulcer disease, a history of alcohol or substance abuse, major hematologic, renal, or hepatic abnormalities, or a recent (within the past six months) stroke or myocardial infarction, or receiving nitrate therapy (Goldstein *et al.*, 1998).

Although the age characteristics of the study population was similar to that of subsequent studies of actual patients (mean age of 61 years), the characteristics of actual sildenafil users differ substantially from those of the subjects who participated in the widely-publicized clinical trials done by the Sildenafil Study Group in 1998. Harrold *et al.* agree, noting that their study cohorts, as well as those of other researchers, were significantly more likely to have concomitant conditions such as hypertension, ischemic heart disease, hyperlipidemia, and diabetes mellitus (2000). Salonia *et al.* report on a study by Padma-Nathan and colleagues regarding the long-term safety results from patients who have been exposed to sildenafil for up to 4.5 years (2003). The study found that the incidence of reported side effects was higher, ranging from 16% to 63% more than those reported in clinical trials. As such, data from the Goldstein *et al.* randomized clinical trials may present a picture that is overly optimistic regarding the effectiveness and tolerability of the treatment, owing to careful patient selection as well as to intensive patient surveillance and monitoring. One of the safety issues with Viagra is that many of the men most likely to use it also have other health issues which were not addressed by Pfizer-sponsored research. This is why Goldstein's trial on primarily healthy men is so misleading. In addition, in the study, sildenafil efficacy was only assessed by using the responses to question 3 (frequency of penetration) and question 4 (maintenance of erections after penetration) of the 15-question

International Index of Erectile Function. Despite the carefully selected, and somewhat idealized group of study subjects, and the fact that transient treatment-related side effects were reported by about 20% of the study participants, including a 2% withdrawal rate due solely to the side effects, sildenafil as well as the study itself were both considered a success (Goldstein *et al.*, 1998; Litwin, 1999).

The large amount of favorable research and publications (fueled by funding from Pfizer) which served as evidence for Viagra's efficacy and safety, undoubtedly served as resources for physicians and to a lesser extent, the public. In fact, Viagra was on the market before many physicians were introduced to the drug and before Pfizer product representatives were trained about the drug and sent into the field (Borzo, 1998). Borzo also notes that in the initial weeks subsequent to Viagra's approval in the United States, physicians had only the package insert to inform them about the drug's indications, counterindications, and interactions with other drugs (1998). It is unlikely that physicians were simply forgotten or abandoned during this crucial period, but instead had Pfizer-sponsored research studies and publications to refer to.

The Issue with Side Effects

Within four months of Goldstein *et al.*'s influential article being published in *The New England Journal of Medicine*, the same journal published seven letters to the editor from physicians bringing forth additional, unexpected side effects of sildenafil as well as shortcomings of the study. In the database, 76 out of the 745 journal articles are dedicated solely to the examination of side effects of the drug, with the instigator published just six months after Viagra's release in the United States by the *American Journal of Ophthalmology*. The article reported the development of unilateral pupil-sparing third nerve palsy in a 56-year old male patient after administration of Viagra, and was followed by many reports of men experiencing eye problems such as changes in the perception of color hue or brightness.

Reports of side effects, while first confined largely to ocular matters, soon began surfacing in various publications. By 1999, an assortment of journals from various medical fields began reporting assorted previously unknown side effects and drug interactions of Viagra. In 1999 alone, nine articles in our database focused solely on side effects of sildenafil, most of which centered on ocular and cardiac functions such as heart attacks. Numerous studies into the side effects caused by the drug were initiated and published in the coming years, including some sponsored by Pfizer. This trend is reflected in our database, where the number of journal articles focusing primarily on the safety and/or side effects of sildenafil, by year, is presented in Figure 2. The number of articles published reached a plateau in 2001 and 2002, with fifteen articles published in each year, after which this number, per year, decrease. The nature of the side effects reported tended to regularly include optic and cardiovascular functions, but also began to include a multitude of symptoms such as nose bleeds, reduced memory function, negative drug interactions and psychosocial changes in patients being examined.

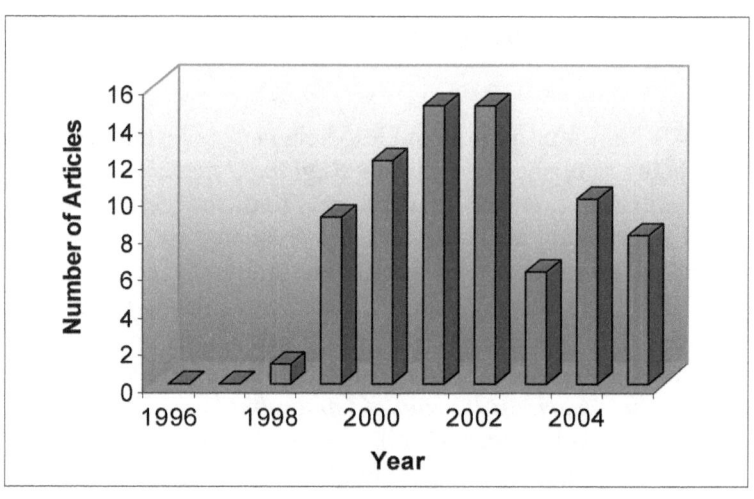

Figure 2 - The number of journal articles in the database focusing primarily on the side effects of sildenafil, by year.

Issues with side effects became more somber as Viagra began being linked with deaths. The drug was linked to 240 deaths reported to the FDA during its first seven months of availability, and to 522 reported deaths after 13 months of availability (Cohen, 2001). In addition, Cohen states that during this time, 1473 major drug reactions were reported to the FDA. Today, Pfizer and the FDA claim that Viagra has caused no more deaths than would be expected among several million men with medical problems commonly associated with erectile dysfunction, like heart disease, high blood pressure, and diabetes, who returned to engaging in sex. Cohen however, remains unconvinced. He compared mortality rates associated with the ED treatment Caverject with those of Viagra, and came up with a higher death rate associated with Viagra. *JAMA* also reported on this issue, and found similar results. Figure 3 illustrates that Viagra is in fact responsible for a ten-fold increase in number of deaths of men per every one million prescriptions, as compared to other modes of treatment for erectile dysfunction.

Deaths of Men Treated for Erectile Dysfunction Through July 8, 1999

Drug	No. of Deaths	No. of Prescriptions, in Millions	Deaths per 1 Million Prescriptions
Caverject (alprostadil for injection)	5	~1.3	4.5
MUSE (transurethral alprostadil)	2	~1.3	1.5
Yocon (yohimbine)	1	~4.0	0.25
Viagra (sildenafil citrate)	564	~11.0	49

Source: John Urquhart, MD

Figure 3 – The number of deaths of men treated for erectile dysfunction based on treatment method for the condition; from *JAMA, 2000; 283:590-593.*

While a November 1998 FDA report stated that only 34% of sildenafil-associated deaths occurred within four to five hours of use (the half life of the compound), reexamination of those data by Cohen reveals that in actuality, 66% (44/67) occurred within this time frame (2001). These findings suggest a clearer sildenafil-fatality causing link. In addition, at this time, the FDA also began requiring that additional warnings about Viagra's use by patients with cardiovascular disease were included in package inserts. However, as physicians do not routinely reread revised package inserts or new editions of the *Physicians' Desk Reference*, these amendments may not necessarily change attitudes and methods of prescribing. The actions of the FDA confirmed earlier concerns by members of the medical profession that sildenafil posed an additional risk to these patients, a risk which was not adequately addressed in initial clinical trials (due to the idealized study group) or initial package inserts. Perhaps some of these deaths were in fact, related to this omission. Cohen agrees by stating that, "What has not been answered is why this vital information was omitted initially, an omission that appears to have cost lives, and what steps the FDA has taken to avoid such potentially harmful oversights in the future". Still remaining unresolved is whether these additional warnings have reduced side effects and deaths in sildenafil patients, since the FDA has not released any subsequent analyses of sildenafil-related side effects.

The propagation of safety concerns and side effects stemming from Viagra use are seen by some as "sporadic reports of side effects published in the literature and sensationalized by the media" and "fear mongering" by others (Jiann *et al.*, 2003). Nonetheless, the findings of research articles, as well as negative media and public responses to the much publicized side effects and deaths resulting from the use of Viagra have undoubtedly increased concern and reduced interest among the public. In their retrospective assessment of the clinical uses and results of sildenafil in the treatment of erectile dysfunction in daily clinical practice in Taiwanese men, Jiann *et al.* found that 39 out of 1290, or 3% of men did not use sildenafil after obtaining it from physicians on solely on account of fear of side effects (2003). Hence, while the efficacy of Viagra has been proven, reports of side effects, whether exaggerated or not, are reducing sales of the drug. Such actions have officials at Pfizer undoubtedly worried, and as such, searching for methods to downplay side effects and their rates.

Through an examination of Pfizer-sponsored research articles, it is apparent that there exists a tendency to downplay, disregard, and find alternate explanations for the side effects experienced by both study participants and patients. For example, in the initial study attempting to determine the efficacy and safety of Viagra, Boolell and colleagues found that six of the twelve subjects, or 50% of the study cohort, reported side effects such as headache, backache, dyspepsia, pelvic musculo-skeletal pain and one case of severe headache (1996). These side effects were characterized as "mild and transient in most cases", a phrase which is repeated three times throughout the paper (1996). In their concluding statement, the researchers affirm that "sildenafil is a well-tolerated and effective oral therapy".

The previously-mentioned article by Goldstein *et al.* (1998) follows similar protocol with regards to the wording employed when describing the side reactions of study participants. During the dose-response study on sildenafil, four men (1% of study cohort) are reported to have stopped taking the drug due to treatment-related side effects (nausea and vomiting in one, leg pain and backache in one, intermittent headache and dyspepsia in one, and headache in one). In addition, in the dose-escalation study, one man is reported to have stopped taking sildenafil due to treatment-related headache and flushing. The authors state that the most frequently reported side effects of sildenafil in the two studies were "transient" and "usually mild" headaches, flushing, dyspepsia, and rhinitis (1998). "Only" one man who experienced flushing and a visual disturbance discontinued treatment. Further, the authors also state that "transient" visual disturbances (changes in the perception of color hue or brightness) were reported by "some" men and that the frequency of these side effects increased with increasing doses of sildenafil, but the symptoms were "usually mild and lasted a few minutes to a few hours after dosing" (1998). The article concludes that "sildenafil treatment was well-tolerated", and that sildenafil's vasodilator properties have "no effect" on heart rate (1998). In practice, this latter conclusion was found to be erroneous and dangerous as a result.

Another observation which emerges from the examination of Pfizer-sponsored literature on Viagra is the renaming of side effects into 'adverse effects'. This new phrase, like the change from impotency to erectile dysfunction, suggests a 'softer', more refined name for unpleasant conditions. This may be due to the fact that while side effects refer to physical experiences or states which occur completely 'on the side', or in addition to the expected results, adverse effects symbolize direct results of the drug, and not mere by-products. In addition, from past experiences, most individuals would most likely relate the phrase 'side effects' with events which would most likely be negative. From this interpretation, adverse effects are indeed more easily accepted, and do not seem as intimidating to the patient as side effects.

Expanding the Market

Maintaining Patients, Maintaining Sales

Eight years after the initial approval of Viagra by the FDA, its efficacy is uncontested, issues regarding side effects have relatively subsided, and more than 110 countries have followed suit by approving the drug for sale (Carson, 2003). However, sales have begun to taper off in North America. In addition, officials and stockholders at Pfizer are likely nervous about the high rate of abandonment of the treatment. A study in the *International Journal of Impotence Research* determined the rate of abandonment of sildenafil therapy and assessed the reasons for abandonment (Klotz, 2005). The study found the rate of abandonment to be 31%, with 45% of these men citing lack of opportunity or desire for sexual intercourse as the main reason for abandonment, and a further 23% reporting that their partners had shown no sexual interest during treatment. Clearly, Viagra will not be sold if partners are not willing to engage in sexual intercourse, regardless of whether or not men actually receive prescriptions for the pills.

A study examining the sexual experience of female partners of men with erectile dysfunction, coauthored by Pfizer-sponsored researchers, was published in September of 2005 in the *Journal of Sexual Medicine*. Perhaps this strategy, of focusing on women's attitudes and catering to their feelings with regards to male erectile dysfunction and its treatment, is the newest method to strengthen Viagra sales. With the drug targeted at middle-aged and older heterosexual men, naturally women in these age categories also serve as an important demographic. This may hint as to why Pfizer's newest television commercials feature middle-aged women singing the famous "good morning" tune.

In 2004, a novel Pfizer-sponsored study was published. In it, Jiann *et al*. assessed "suboptimal" use of Viagra in a subset of ED patients in whom there was no response to sildenafil, as well as the feasibility of successfully "rechallenging nonresponding patients" (2004). In the study, Taiwanese men who had claimed a poor response to sildenafil were assessed with regards to 'inappropriate' use of the drug. The authors claim that thorough instruction and re-instruction by the prescribing physician, and the administration of a minimum of four separate doses of 100mg sildenafil (the highest dose possible) are considered adequate "attempts" at treatment of erectile dysfunction. In addition, the authors state that patients must be educated and followed regularly in order to obtain optimal results with regards to sildenafil rechallenging. A lack of any of these factors constitutes an inadequate attempt, and the patient is seen as being in need of "correction". According to the article, it is clearly the patient who is considered to be at fault for a lack of response to sildenafil if he does not experience positive results prior to administrating less than four maximal doses!

Another new concept aimed at increasing sales of Viagra involves taking Viagra on a daily basis in order to prevent erectile dysfunction in the future. The *British Medical Journal* published a news article in 2003 regarding events which unfolded before a large audience of primary care physicians and others attending a Pfizer-sponsored sexual function update at New York University. Dr. Goldstein, a panelist, stated that he had "hundreds of men" using Viagra for prevention: "If you would like to be sexually active in five years' time, take a quarter of a pill a night; we have data that show that will facilitate and prolong nocturnal erections." (Moynihan, 2003). The comments were made during a panel session, in response to a question regarding the role of sildenafil-type drugs in daily usage from another urologist, who told the meeting "I'm a strong advocate of proactive prevention as well." Although the question appeared to be one of several collected from the audience on pieces of paper before the session started, this second urologist, Dr. Bar-Chama of the Mount Sinai School of Medicine, said later that he had thrown in the question himself. When asked whether either doctor had ties to Pfizer, Dr. Bar-Chama replied, "On occasion I speak for them".

Perhaps this was an isolated event, but most likely, Pfizer, through their multitude of sponsored researchers and physicians, is attempting to increase sales through yet another means. There is currently no available research on the topic of daily prevention strategies through the use of sildenafil. Hence, this method of 'daily prevention' seems neither proven, nor indicated, and as such, may indeed be harmful.

Novel Uses for Viagra

Research in the field of erectile function and dysfunction has continued to expand rapidly. Based on the information available, some directions for future erectile dysfunction therapies can be identified (Andersson and Hedlund, 2002). The first direction is improvement of current therapeutic principles, such as with the newly introduced second generation of orally active PDE inhibitors. Another direction is the development and administration of combinations of existing therapeutic agents such as combinations of apomorphine and sildenafil. These methods seem attractive and may have a therapeutic potential in patients not responding satisfactorily to single-drug treatments in the future (Andersson and Hedlund, 2002).

Since the development of Viagra, the industry has been scrambling to develop an equivalent aid for women, encouraged by significant female demand. Women have spent a minimum of $200 million for clinically proven phoney products such as Avlimil, which was earlier on touted as a natural female enhancement pill (Read, 2005). Thus far, however, the attempts to produce veritable female products have been unsuccessful while at least three are readily available for men. So why has it taken this long to develop a drug which specifically addresses women's sexual needs? Sex therapist Judith Seifer writes that regarding sexual function, "we're about fifteen to twenty years behind in our studies with women" compared to men (2004). Ellison writes that the FDA is under increasing pressure from congress and the public in order to clamp down on safety standards after a number of high-profile drugs, including Viagra, Vioxx and menopausal hormone therapies have been shown to cause serious side effects, which manufacturers were often aware of beforehand (Raja and Nayak, 2004).

The main reason why the market has been slow to develop is most likely due to social taboos regarding women's sexualities, as well as what is assumed to be women's more nuanced and complicated sexual desires, which are more difficult to quantify physically, and as a result, to develop a pharmaceutical treatment for. Although an erect penis functions as an obvious sign of whether a man is 'turned on', it does not necessarily mean that men's sexual desires are less "nuanced" or complicated. However, sexual dysfunction in men has been largely defined as the inability to get or maintain an erection. In contrast, there is no precise definition for female sexual function. It includes disorders of desire and arousal, difficulties in achieving orgasm as well as pain tied to intercourse. Hence, sexual function in women can not be analyzed and treated simply as a physical issue, as erectile dysfunction largely was. Nonetheless, pharmaceutical companies are keen to capitalize on this market.

The term 'female sexual disorder' (FSD) first appeared in the medical literature as a symptom of dyspareunia in 1975. This term serves as an umbrella term containing all of the possible 'symptoms' associated with inadequate female sexual function. Due to the lack of precise definitions in this field, inadequacy is a very subjective term. In addition, the related term, 'female sexual arousal disorder' (FSAD) first appeared in the medical literature in 1990. Not surprisingly, with the increased attention to, and cultural acceptance of erectile dysfunction, Pfizer initiated studies to improve their understanding of female sexual function, as well as to perhaps medicalize and develop a market for both female conditions. An examination of medical journal articles on the topic, as with those on erectile dysfunction, reveals major links between Pfizer and many researchers in the field. In fact, many of these researchers are the same ones who first exposed and publicized erectile dysfunction: The Sildenafil Study Group, Dr. Goldstein and Dr. Althof, the editor-in-chief of The International Journal of Impotence Research, to name a few.

With the appropriate terms in place, it is highly likely that Pfizer has begun to explore this highly lucrative market with regards to possible treatments and associated marketing strategies. It would of no surprise to witness the rise of a new 'female Viagra' via the use of the same principles which were employed to popularize Viagra itself. It is interesting to note that the *International Journal of Impotence Research* has continued to employ the term 'impotence' in its title despite Pfizer's and arguably the general scientific field's, preference for the term erectile dysfunction. With these new developments in the realm of female sexuality, it can be ventured that the journal will soon adopt a name which relates equally to both sexes, and thus can serve as a vehicle to promote and defend the new treatment.

Currently however, the fastest-growing fields for sildenafil treatment lie in various medical disciplines and disorders. For example, small doses of sildenafil may be a useful adjunct to inhaled iloprost, the currently preferred medication in the management of pulmonary hypertension. In gastrointestinal disorders, sildenafil also exerts several effects which might be of clinical relevance. In patients with heart failure, endothelial dysfunction is influenced by the PDE 5 inhibitor and exercise capacity may be improved. Furthermore, in the treatment of Reynaud's phenomenon, a disease without any highly effective medical treatment options yet, first observations with sildenafil appear to be promising (Cremers and Bohm, 2003).

Research into sildenafil as a compound, and not as a treatment for erectile dysfunction first appeared in our database in 1999 in an *American Journal of Cardiology* article. The article featured an exploration of the effects of sildenafil on tissue cyclic nucleotides, platelet function, and the contractile responses of heart tissues *in vitro*. As is apparent from Figure 4, the number of journal articles employing sildenafil as a treatment for a medical condition other than erectile dysfunction has steadily increased from 1999 onwards. This trend is also present with regards to the fraction of articles employing sildenafil as a treatment for a medical condition other than erectile dysfunction, as a proportion of the total number of articles published per year. Out of 745 articles in the database, 441, or 59% employ sildenafil solely as a treatment for erectile dysfunction, the remainder focusing on the treatment of such wide-ranging conditions as penile sarcoma, the improvement of endometrial thickness, esophageal motor function and lung fibrosis.

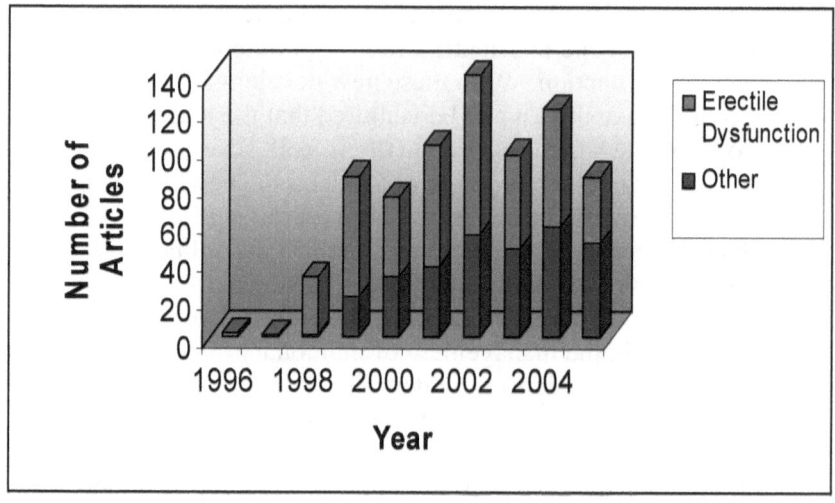

Figure 4 - The number of journal articles in the database grouped by the primary clinical use of sildenafil in the article, by year.

In the database, the most frequent non-erectile dysfunction related condition for which sildenafil is employed is pulmonary hypertension. In their literature review of studies investigating the use of sildenafil for the treatment of this condition, Lee *et al.* concluded that sildenafil reduced pulmonary arterial hypertension and pulmonary vascular resistance and tended to increase cardiac output compared with baseline (2005). As such, the authors conclude that sildenafil is a promising and well-tolerated agent for the management of pulmonary hypertension. As is expected, in many of these articles, especially those in cardiology journals, Pfizer is a sponsor of the research, researcher, or both. Undoubtedly, if sildenafil can be used by additional subsets of patients, Pfizer stands to gain larger markets and increased sales.

Influencing International Patients and Physicians

There exists a multitude of literature, most of which is published in the *International Journal of Impotence Research*, openly attempting to evaluate the efficacy, safety, and tolerability of Viagra in men with erectile dysfunction of various etiologies (organic, psychogenic, or mixed). One subset in this group of journal articles however, is solely interested in examining the efficacy of Viagra on men of different races. These articles report on studies which are most often performed by researchers in the country of interest, and financially supported by Pfizer. One common link between many of the studies is their tendency to take on a very similar form, often being randomized, double-blind, and placebo-controlled. Regardless of study population or ethnicity, all studies report very similar results to those of clinical trails conducted in Western countries.

One large scale study in this cohort was performed by the *Asian Sildenafil Efficacy and Safety Study Group (ASSESS – 1)* on Malaysian, Singaporean, and Filipino men with erectile dysfunction. In the introduction, the article states that in Asia, the exact prevalence of ED is unknown. It continues that "if comparable to the United States, where ED of some degree has been estimated to affect 52% of men 40 to 70 years of age, the scale of the problem in Asian countries is probably very large" (Tan *et al.*, 2000). As expected, the article concludes that "the results in the trial in Asian men are similar to those reported in Western populations".

Why are all of these studies performed at the expense of time, energy and substantial funding? All of them conclude that the findings with regards to the efficacy, safety and tolerability of Viagra in men of a certain ethnic group are similar to those reported in Western populations. These factors, especially Viagra's efficacy, are well-known, have already been proven through scientific research, and are not expected to alter with regards to ethnicity. Why do researchers, and ultimately individuals at Pfizer believe that Viagra will work differently on men of different races? Seldom are medications tested in this manner. The majority of countries do not require a study to be conducted on a specific population in order to approve a new pharmaceutical agent. Even if this was the case, the specific subset of studies in question most often deals with a grouping of men from surrounding regions, not solely one country. There is a lot that can be written about the issue of race and biology with regards to these undertakings. One important point to note is that such studies reify 'race', in effect confirming the idea, for which there is no biological basis, that there is something inherently different between racial or ethnic groups.

It can be argued that sex for recreation is important to our culture. Perhaps this is not so to individuals living in other cultures and other countries. These individuals are undoubtedly seen through the lens of economics, and as such, as entire markets, by Pfizer. Hence, the articles may be attempting to condition and prepare individuals outside of North America and Europe. By creating interest in, and instilling the importance of this form of sex in 'others', it seems that Pfizer may be attempting to create a market for erectile dysfunction (and possibly FSD). Are individuals of these countries perhaps being conditioned to consider erectile dysfunction as a 'big deal' and a bodily condition in need of treatment? In this subset of articles, researchers often address cultural implications of arousal and sexuality by referring to their usage of the *International Index of Erectile Function*. In theory, the index is noted to be "cross-culturally valid", and although it is available in ten languages, there is no evidence of cultural implications or sensitivity to different interpretations of sex-related issues and lifestyles (Rosen *et al.*, 1997). In addition, since most often the entire index is not employed, but only certain questions, its validity is questionable.

As mentioned above, it is hypothesized that these studies, rather than focusing solely on the efficacy and side effects associated with Viagra as they overtly claim to do, are part of a larger agenda to increase sales of the drug. Perhaps, as

Loe hypothesized, Pfizer is attempting to create markets based on cultural need (Loe, 2004). Or, perhaps Pfizer is attempting to make physicians at ease with the prescription of Viagra to their respective populations. If any uncertainty exists about whether or not Viagra is safe to use on a certain population, these studies satisfy. They often promote the fact that patients are not overly troubled with the price of the drug (half of a sample of 234 men were willing to pay 25 Euro per month for ED treatment, and 8% were willing to pay any amount), but rather curious about its efficacy and apprehensive about possible side effects, as may be their physicians (Klotz *et al.*, 2004). If physicians are reassured that a study on individuals of a particular ethnic group found Viagra to be well-tolerated, not only can they pass this information on to patients, but will also be more likely to issue prescriptions. Hence, these studies likely function to fill a more covert role, and in the process draw attention to and support the medicalization of a problem which may not be necessarily seen as such in certain parts of the world.

Conclusion

The Viagra phenomenon, at its most basic level of analysis, has changed our understanding of sex in North America, and increasingly around the world as well. Normal sex now means sex on demand, sex for everyone, and sex for life (Loe, 2004).

Although treatment of erectile dysfunction with Viagra has been shown to significantly improve quality of life parameters related to sexual dysfunction and mental health (Salonia *et al.*, 2003), one argument among many others, is that the emergence of Viagra has only intensified our otherwise sexualized society in which sexual health and pleasure are endlessly promoted and appear to be the keys to life itself (Loe, 2004). Beyond these shifts in definitions however, the phenomenon has also impacted our broader ideas about health, aging, and masculinity.

In 1993, the NIH Consensus Development Panel on Impotence reported that "appallingly little is known about the prevalence of erectile dysfunction in the United States and how this prevalence varies according to individual characteristics (age, race, ethnicity, socioeconomic status, and concomitant diseases and condition)". Due to the plethora of interest and research on ED and ED-related matters that has occurred within the past ten years, most of which is largely due to the development of Viagra, the universe of knowledge regarding erectile dysfunction is increasing. As a result, researchers, and ultimately patients, have benefited from these events to some degree.

At a deeper level of analysis, the case of Viagra exposes how the construction and dissemination of 'facts' can be undertaken by corporations, how diagnostic expansion can work in partnership with market expansion, and how medicalization can become synonymous with increased profits. This study, although focused specifically on Viagra, poses new, crucial questions about the intersection of science, treatment and capital. Arguably, pharmaceutical manufacturers have become "deeply enmeshed" in the process that determines which drugs we use and when, why and how we use them (Greider, 2003; Cockburn and Henderson, 1996). Critics, including Marcia Angell, former editor of the *New England Journal of Medicine*, have argued that the shift from the academic to the commercial sector has given the industry too much control over clinical drug-trial design, data analysis and publishing (Angell, 2000).

My findings, taken primarily from the medical literature, suggest that the corporate infiltration of medicine, often done through the manipulation of scientific research, is prevalent. Clearly, current safeguards towards unbiased medical information are lacking. These findings also suggest that expert opinions may be 'cloaked' marketing. Hence, funding may supply scientific support for a drug which can then appease regulators, and allow sales. What do these findings say about medicine and the medical profession today? It appears that through key funding of certain researchers and publications, Pfizer made erectile dysfunction into a legitimate disease and Viagra into a legitimate treatment without necessarily going through the most prestigious channels. Obviously, some medical professionals played a role, but not the general medical profession as a whole.

From my analysis, I argue that Relman's medical-industrial complex is supported by this case, and that unfortunately this does not bode well for our society, and increasingly for other societies worldwide. The case of Pfizer's Viagra illustrates that today's market-oriented, profit-driven health care industry can be influenced from the top down simply due to funding of the right researchers and marketing teams. This results in potentially fatal results for some patients when compounds are not screened properly prior to approval by regulatory agencies, when side effects are overshadowed by manufacturers, and when paid researchers are keen on increasing sales of drugs through methods which are not properly assessed. It appears that we are distancing ourselves from a completely unbiased, patient-centered state of medicine which many believe still exists. We are also quick to purchase treatments for what Figueras et al. call "induced" health needs (2002). Since these findings result solely from the case of one 'blockbuster' drug, many similar issues likely exist with compounds in similar categories, as well as with those which serve completely different subsets of conditions.

There are improvements that must be made in order to improve the quality of our current medical system. While there does exist evidence of corporate infiltration at every level of the creation and dissemination of scientific information, ultimately users of this information hold great capacity. Both prescribers and users of Viagra, and of similar pharmaceuticals, decide whether to support such corporations. Although it is naïve to state that complete abstinence from pharmaceuticals, many of which we are often reliant on, is possible; researching, questioning and speaking out against certain practices of pharmaceutical companies and physicians, is not.

Literature Cited

Andersson, KE, Hedlund, P. (2002). New directions for erectile dysfunction therapies. *International Journal of Impotence Research,* 14 Suppl. 1, S82-S92.

Angell, M. (2000). Is academic medicine for sale? *New England Journal of Medicine, 342,* 1516-1518.

Anonymous. (1999). Canadians traveled "Viagra Highway" after drug's release delayed. *Canadian Medical Association Journal, 160,* 769.

Boolell, M., Gepi-Attee, S., Gingell, J.C., & Allen, M.J. (1996). Sildenafil, a novel effective oral therapy for male erectile dysfunction. *British Journal of Urology, 78,* 257-261.

Borzo, G. (1998). Viagra raises demand, questions. *American Medical News, 41*(21), 25-26.

Braun, M., Wassmer, G., Klotz, T., and Reifenrath, B. (2000). Epidemiology of erectile dysfunction: results of the 'Cologne Male Survey'. *International Journal of Impotence Research, 12*(6), 305-311.

Carpiano, R. M. (2001). Passive medicalization: the case of Viagra and erectile dysfunction. *Sociological Spectrum, 21,* 441-450.

Carson III, C.C. (2003).Sildenafil: a 4-year update in the treatment of 20 million erectile dysfunction patients. *Current Urology Reports. 4,* 488-496.

Cockburn, I., & Henderson, R. (1996). Public-private interaction in pharmaceutical research. *Proceedings of the National Academy of Science of the USA, 93,* 12725-12730.

Cohen, J.S. (2001).Is the sildenafil product information adequate to facilitate informed therapeutic decisions? *The Annals of Pharmacotherapy. 35,* 337-342.

Conrad, P., and Leiter, V. (2004). Medicalization, markets and consumers. *Journal of Health and Social Behavior, 45,* 158-176.

Conrad, P., & Potter, D. (2000). From hyperactive children to ADHD adults: observations on the expansion of medical categories. *Social Problems, 47,* 559-582.

Cremers, B. & Bohm, M (2003).Non Erectile Dysfunction Application of Sildenafil. *Herz. 28,* 325-333.

Feldman, H., Goldstein I., Hatzichristou D., and Krane, R. (1994). Impotence and its medical and psychosocial correlates: results of the Massachusetts Male Aging Study. *The Journal of Urology, 151,* 54-61.

Figueras, A., Vasquez S., Arnau J., and Laporte, J. (2002). Health needs, drug registration and control in less developed countries - the Peruvian case. *Pharmacoepidemiology and Drug Safety, 11,* 63-64.

Goldstein, I., Lue, T.F., Padma-Nathan, H., Rosen, R.C., Steers, W.D., & Wicker, P.A. (1998). Oral sildenafil in the treatment of erectile dysfunction. *The New England Journal of Medicine, 338,* 1397-1404.

Greider, K. (2003). *The Big Fix: how the pharmaceutical industry rips off Americans.* New York: Public Affairs Publishers.

Harrold, L.R., Gurwitz, J.H., Field, T.S., Andrade, S.E., Fish, L.S., & Jarry, P.D. (2000). The diffusion of a novel therapy into clinical practice. *160,* 3401-3405.

Jarow, J., Kloner, R., & Holmes, A. (1998). *Viagra: how the miracle drug happened & what it can do for you!* : M. Evans and Co.

Jiann, B.P., Yu, C.C., Su, C.C., & Huang, J.K. (2004). Rechallenge prior sildenafil nonresponders. *International Journal of Impotence Research, 16,* 64-68.

Kinsey, A., Pomeroy, W., & Martin, C. (1948). *Sex-ual behavior in the human male.* Philadelphia, PA: W.B. Saunders.

Klotz, T, Mathers, M, Klotz, R, & Sommer, F (2004). Why do patients with erectile dysfunction abandon effective therapy with sildenafil (Viagra)? *International Journal of Impotence Research, 17,* 2-4.

Latour, B. (1987). Science in Action. Cambridge, Massachusetts: Harvard University Press.

Lee, A., Chiao, T., & Tsang M. (2005). Sildenafil for pulmonary hypertension. *The Annals of Pharmacotherapy, 39*(5), 869-884.

Levine, S. (1976). Marital sexual dysfunction: erectile dysfunction. *Annals of Internal Medicine, 85*(3), 342-351.

Litwin, M. S. (1999). Urology. *The Journal of the American Medical Association, 281*(6), 495-496.

Loe, M. (2004). *The Rise of Viagra: how the little blue pill changed sex in America.* New York: New York University Press.

Lupton, D. (2003). *Medicine as culture - illness, disease and the body in western societies.* 2nd ed. London: Sage Publications.

Maguire, P. (1999). How direct-to-consumer advertising is putting the squeeze on physicians. *American College of Physicians' Observer*, published by the American College of Physicians and the American Society of Internal Medicine in March, 1999.

Mintzes, B., Barer M. L., Kravitz R. L., Bassett, K., Lexchin, J., Kazanjian, A., Evans, R.G., Pan, R., and Marion, S.A. (2003). How does direct-to-consumer advertising (DTCA) affect prescribing? A survey in primary care environments with and without legal DTCA. *Canadian Medical Association Journal, 169*(5), 405-412.

Moynihan, R. (2003).Urologist recommends daily Viagra to prevent impotence. *British Medical Journal. 326*, 9.

National Consensus Development Panel on Impotence. (1993). Impotence. *Journal of the American Medical Association, 270*, 83-90.

Pfizer Pharmaceuticals Website, (2005). About ED. Retrieved Mar. 30, 2006, from ED: A Common Issue Web site: http://www.viagra.com/faqs/faqs2.asp.

Potts, A., Grace, V., Gavey, N., and Vares, T., (2004). "Viagra stories": challenging erectile dysfunction. *Social Science and Medicine, 59*(3), 489-499.

Raja, S., & Nayak, S. (2004). Sildenafil: emerging cardiovascular indications. *The Annals of Thoracic Surgery, 78*(4), 1496-1506.

Relman, A.S. (1991).The health care industry: where is it taking us? *New England Journal of Medicine. 325*, 854-859.

Read, J., (1995). *Female sexual dysfunction.* International Review of Psychiatry, 7(2), 175-182.

Rosen, R. C. (1998). Sildenafil: medical advance or media event? *The Lancet, 351*(9116), 1599-1600.

Salonia, A., Rigatti, P., & Montorsi F. (2003). Sildenafil in ED: a critical review. *Current Medical Research & Opinion, 19*(4), 241-262.

Soderling, S.H., & Beavo, J.A. (2000). Regulation of cAMP and cGMP signaling: new phosphodiesterases and new functions. *Current Opinion in Cell Biology. 12*, 174-179.

Stipp, D., & Moore, A. H. (1998). Impotence is a much bigger problem than doctors used to think, and new pills are on the way to treat it. *Fortune, 137*(5).

Tan, H.M., Moh, C.L., Mendoza, J.B., Gana, T., Albano, G.J., & de la Cruz, R. (2000). Asian sildenafil efficacy and safety study (ASSESS-1): a double-blind, placebo-controlled, flexible-dose study of oral sildenafil in Malaysian, Singaporean, and Filipino men with erectile dysfunction. Urology, *56*, 635-640.

Tiefer, L. (1994). The medicalization of impotence: normalizing phallocentrism. *Gender & Society, 8*, 363-377.

Tsao, A., (2004). Can Intrinsa be a Viagra for women? *New Scientist,* 184(2477), 4.

United States National Library of Medicine - National Institutes of Health (NIH) Website, (2005). Bibliographic services division. Retrieved Mar. 24, 2006, from Abridged Index Medicus (AIM) Journal Titles Web site: http://www.nlm.nih.gov/bsd/aim.html.

United States Food and Drug Administration Website, (1998). FDA approves impotence pill, Viagra. Retrieved Mar. 10, 2006, from http://www.fda.gov/bbs/topics/ANSWERS/ANS00857.html.

www.ingramcontent.com/pod-product-compliance
Lightning Source LLC
Chambersburg PA
CBHW021252280526
45784CB00005B/2337